Jack Binns to the Rescue!

The Story
of the
First Rescue at Open Sea
Coordinated by Wireless

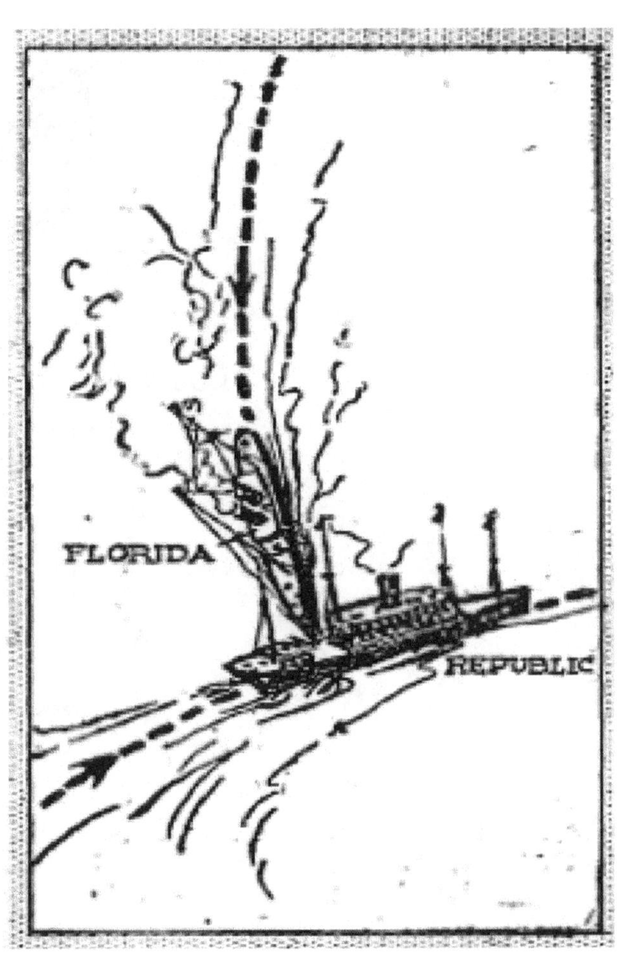

Virginia Utermohlen Lovelace

VUBooks

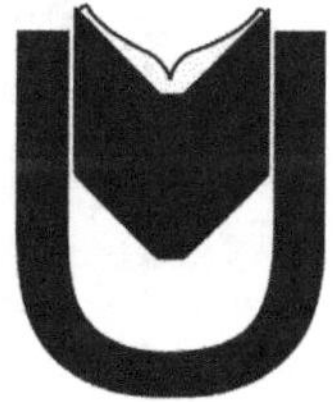

Books with a Point of View
Chelmsford MA

Copyright © 2021 Virginia Utermohlen Lovelace MD

ISBN: 9798613877300

Cover: Photo by Ivana Cajina on Unsplash.com

FOREWORD

CQD! Here MKC! Shipwrecked!

In 1909, three years before the sinking of the *Titanic*, the world watched as the first rescue at open see coordinated by wireless unfolded. More than 1500 lives were at stake.

Jack Binns, the Marconi wireless operator, sent out the cry for help and guided in the rescue ships—and in the process became an international hero.

More than a hundred years later, the honor has fallen to me, his granddaughter, to put together Jack Binns' story, from his birth in an English Poorhouse, to his abandonment by both father and mother, to the horrific accident that nearly took his life but gave him the chance to study and learn and become adept at wireless communication, to the shipwreck and its aftermath, and its consequences for the *Titanic* disaster.

To bring you this story, I've pulled together accounts and images, as well as Binns' own words. I've set out his words in Garamond, so you can distinguish them from my words and enjoy his very own take on these events.

Here's the story!

Virginia Utermohlen Lovelace MD
Chelmsford MA
2021

I would like to dedicate this effort to my grandfather Jack Binns,
my "Binnsy,"
my inspiration and support through my childhood and youth,
whose love, laughter, and wisdom
inspire me still.

TABLE OF CONTENTS

Locations in Great Britain mentioned in Jack Binns' story.

John Robinson "Jack" Binns was born in the Union Poorhouse in Brigg, Lincolnshire, on September 16th, 1884. His mother was an unmarried servant, who gave her firstborn to her family to raise, while she moved to Grimsby.

Not long after he was born, his uncles and widowed grandmother took Jack to live with them in Peterborough, where his uncle William worked as a tailor and his uncle Walter labored on the railroad. They told him that he was orphaned, and he never met his mother again.

At age 13, after quickly completing the required elementary school, Binns went to work to help his family out.

His first job: messenger at the Great Eastern Railway Telegraph Office, a perfect job for someone fascinated by electricity and communication.

As he said:

Binns at age 2.

"The year 1894, when I was 10 years old, brought a marvel for me during the annual agricultural show in that small city—which was also the marketing center of an agricultural area. It was a talking machine. I remember putting the stethoscope-like tubing into my ears—after parting with the tremendous sum of sixpence for the privilege—and listening in amazement to a recitation of the "Lord's Prayer."

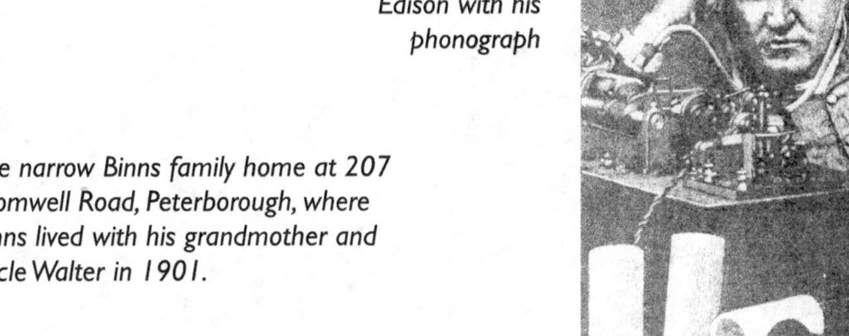

Edison with his phonograph

The narrow Binns family home at 207 Cromwell Road, Peterborough, where Binns lived with his grandmother and uncle Walter in 1901.

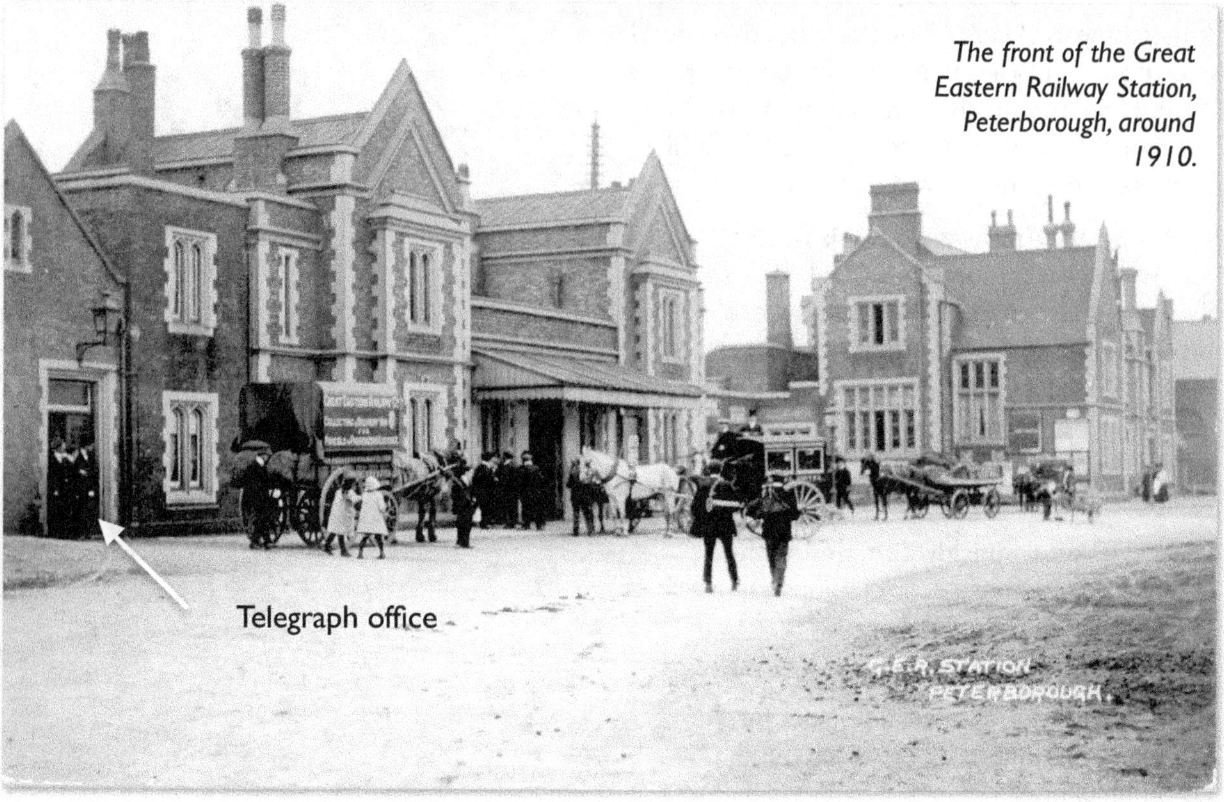

The front of the Great Eastern Railway Station, Peterborough, around 1910.

The marshaling yards of the Great Eastern Railway, where Binns legs were nearly severed by a moving train.

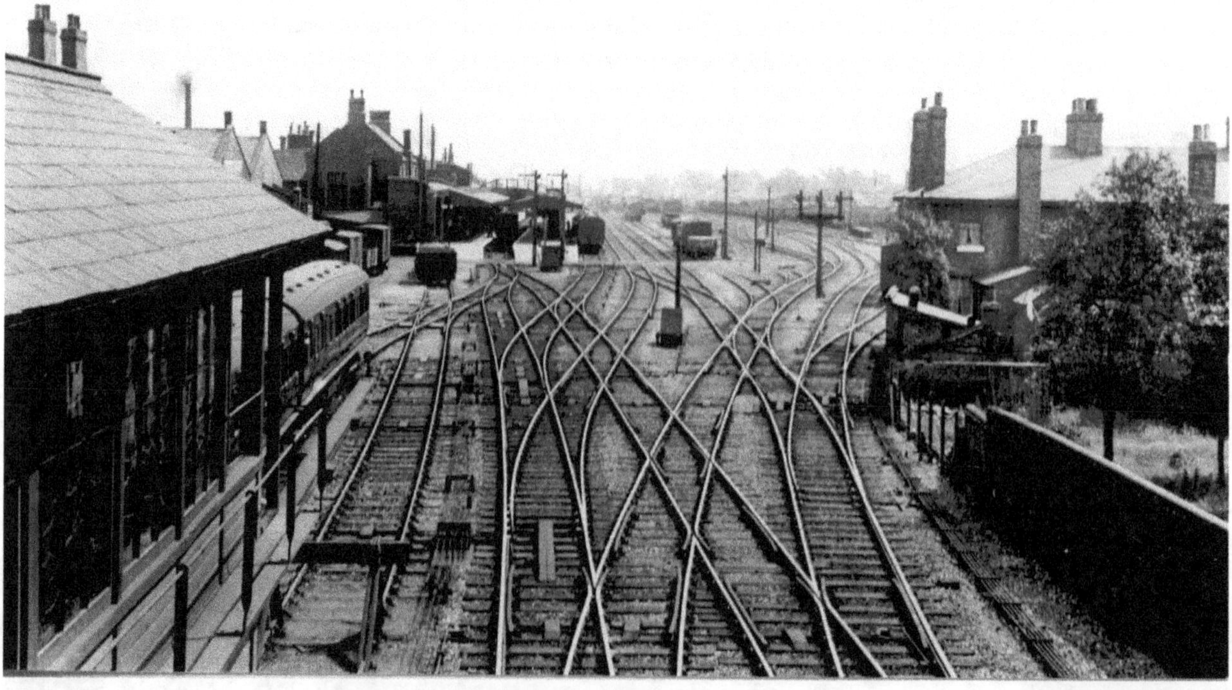

THE FIRST ACCIDENT!

On December 8th, 1898, unknown to me, a steam engine was hitched to one of the strings of cars. Late that afternoon I was given a message to deliver to the other side of the yards. The shortest way was through the freight cars and platform. So, as I had done on so many previous occasions, I ducked under the cars and was on my way. On the return I retraced my steps. As I was about to clear the track, the engine started up and the last coupled car caught my right foot. As I fell my left leg was thrown across the track. The freight car passed over my right foot and cut across my left leg midway between the knee and ankle.

Fortunately I had sense enough not to move, but I was told afterwards that I let out the most bloodcurdling yell any of the men in the yard had ever heard....

When Binns arrived at the hospital the doctors thought he had a slim chance to live so they chose not to amputate his legs. Instead he spent year in traction, barely able to move more than 12 inches back and forth. It took his wounds over two years to heal...

...In one respect the enforced immobilization was a blessing. It enabled me to read and study without interruption—except for the daily lectures and dressings. The doctors were justifiably proud of their work in saving my legs, and also in defeating the serious infection that caused the extremely high fever and threatened my life. As a result I was their pet patient. They readily satisfied my craving for books, and had them brought from the public library, as well as the library in the infirmary. Whenever their time allowed the doctors would sit on the edge of my bed chatting and answering my questions.

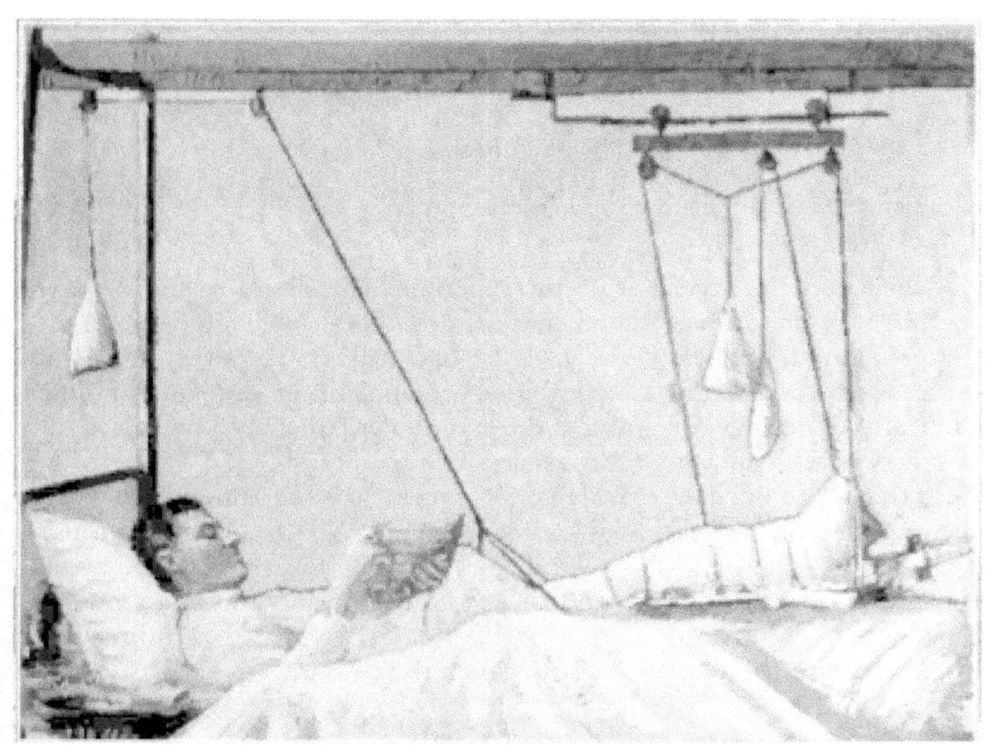

The type of traction system Binns lay in for a year—both of his legs were in traction.

Once I was able to negotiate the distance between my home and the telegraph office, I went back to work on my crutches. This time, however, it was as a junior operator. For more than six months I traversed the mile and a half between home and office on crutches each day every week day, and put in twelve hours work in between—except on the one day each week when I had to go to the hospital and have the unhealed wound on my leg treated. It was another year before I was able to get around with only a cane.

By age 17, in 1901, Binns became senior officer second in charge at the Colchester offices, a center for telegraph communication between England and the entire world!

At Colchester Binns learned how to modify and repair telegraph equipment, knowledge that would become useful during the Republic-Florida disaster!

Connections of the Eastern Telegraph Company, 1901.

In various conversations I had with the heads of the telegraphic department of the railway company, I got the impression that I was secure in my job for life. I sensed that this consideration derived from a feeling of responsibility over the orders which had sent me across dangerous tracks to deliver messages. Whatever the cause, I had no great hankering for that kind of security— especially as I looked around and saw the hopeless hangdog looks which seemed to have been permanently frozen into the faces of my superiors. The desire to move on to greener pastures was strong within me.

I had become adept on every type of telegraphic equipment in use at that time, with the exception of that used on cables and in wireless communication ... I wrote to the Marconi Company to ascertain whether there was any possibility of securing a position with them. My enquiry was timely, and by the end of 1904 I was in the Marconi school at Seaforth, near Liverpool, undergoing instruction in the new art.

BINNS BECOMES A MARCONI MAN

Binns' extensive training and study meant that he could complete the wireless telegraphy course at the Marconi School in Seaforth, near Liverpool, in 3 months, instead of the usual 6 to 10 months!

Marconi was busy equipping ships with his wireless apparatus, so Binns soon found himself sailing across the Atlantic, and even on North Sea cruises.

In July 1906, as part of his research work for Marconi on radio transmission in the Arctic, Binns sent out the first wireless message into the Arctic skies…

"While the midnight sun shines down upon the Northern seas, the steamship Blücher greets old Arcticus with the first radio message ever transmitted within his magic and mysterious circle."

…but there was nobody to answer!

Young Binns in his Marconi cap.

Kaiser Wilhelm der Grosse

Kaiserin Augusta Viktoria

Grosser Kürfurst

In 1906 Marconi equipped German ships to receive messages from both Poldhu in Cornwall, England and Cape Cod, US, making daily newspapers with headlines available for the first time.

The news items were received by the "Marconi Man," then composed and sent down to the ship's printshop.

Another duty of the Marconi Man: franking postage!

This postcard was undoubtedly franked by Binns on a cruise to the Arctic Circle aboard the *Blücher.*

Guglielmo Marconi (1874-19370 was an Irish-Italian nobleman and one of the many pioneers of wireless radiotelegraphy who developed the science and the art at the end of the 19th and the beginning of the 20th centuries.

He gave the world the first practical wireless radiotelegraphy system capable of spanning the Atlantic Ocean, and as a businessman he developed the world's first successful ship-to-ship and ship-to-shore wireless communication system.

In 1909, in part in recognition of the value of this system following the Republic Florida collision, he was awarded the Nobel Prize in Physics together with another pioneer, Karl Ferdinand Braun.

Guglielmo Marconi circa 1903

A "radio shack" on the deck of a passenger ship. These structures were added on to ships that did not have a dedicated radio room.

This photo appeared in a number of publications either as it is here or reversed, purporting to be a photo of the radio shack on the deck of the Republic.

Inside a shipboard radio shack.
Note the bed on the right.
Legend added, courtesy Russ Kleiman.

A. Sending aerial
B. Electrical box
C. "Jigger" for tuning aerial
D. Leyden jars

E. Empty tape reel
F. Key
G. Coherers
H. Receiving aerial

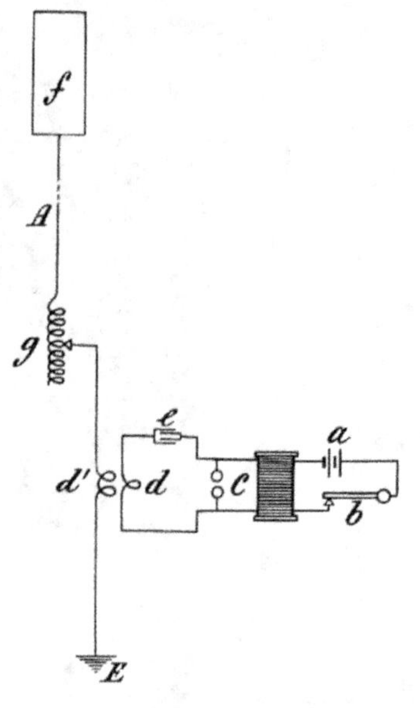

Diagram showing the elements of the wireless sending apparatus, from Marconi's British patent 7777, granted in 1901:

a: the battery; in the case of a ship this could also be the energy provided by the ship's own electricity from a dynamo;

b: the sending key

C: the induction coil and spark gap;

d and d': the coils transmitting current to the aerial f;

e: the 'capacitance' provided by Leyden jars;

E: ground

g: tuning coil

A: wire to the aerial;

f: the aerial

HOW THE MARCONI WIRELESS SYSTEM WORKS

he Marconi wireless system is based on the principle that a current moving back and forth in a wire will radiate its energy into the surrounding air in the form of radio waves. This energy in turn can cause a similar current to be produced in a wire some distance away.

An electric current in a wire comes about when very tiny negatively charged particles, called electrons, moving through a wire. These electrons are attracted to positively charged particles in a wire, so if you make one part of the wire more positively charged, the electrons will move towards that part, creating a current.

The electrons' movement can either be in one direction, called a direct current; or it can go back and forth rapidly if you turn the positive end of the wire on and off, giving you an oscillating or alternating current.

A battery is designed to separate electrons linked to one pole from positively charged particles linked to a second pole. When a wire outside the battery connects the two poles the electrons can move through the wire towards the positive pole, so you have a current going through the wire.

The battery (a) in the diagram is connected by a wire going from one pole to the key (b). The key is connected to the other pole of the battery through the induction coil (C). When the key is down, the circuit is complete, so a current can move through from one pole of the battery to the other. When the key is lifted, the circuit is broken and no current flows.

The induction coil is actually made up of two separate coils of wire. One of these wires is connected to the battery and the key, and the other is coiled around the first wire, but doesn't touch it. When the key is pressed down and a current moves through the first coil, it induces a current to move through the second coil, hence the name of the apparatus.

The circuit containing the induction coil is constructed in such a way that that together with the Leyden jars, the direct current flowing through the first circuit through the key is transformed into an oscillating current. By oscillating the current can build up sufficient pressure, called voltage, to cause the electrons to leap across the gap between the ends of two wires, the spark gap, sending out energy in the form of light —a spark. The spark completes the circuit to the coil d, and sends the oscillating current through the aerial wire to the antenna (f).

The energy of the electrons oscillating back and forth in the antenna is transformed into radio waves that radiate out from the antenna wire. The energy of a radio wave drops off as you move away from the antenna, because it gets transferred to the molecules of air (for example) surrounding the antenna. You need a great deal of energy to send radio waves over great distances. The higher the voltage created by the combination of Leyden jars and induction coil with its spark gap, the further the radio waves can travel.

At the receiving end, the energy is picked up by an antenna, and the current created in that antenna can be guided down by wires to a receiving system. The coherer, which is the receiving system developed by Marconi, contains a tube with metal filings in the middle and wires coming into the time at each end. When a current of sufficient energy comes through the antenna, the filings actually lift up and make a bridge between wires at each end of the tube. This completes a circuit in the receiving system, that ultimately is translated either into dots and dashes inked onto a moving tape, or into sound in headphones.

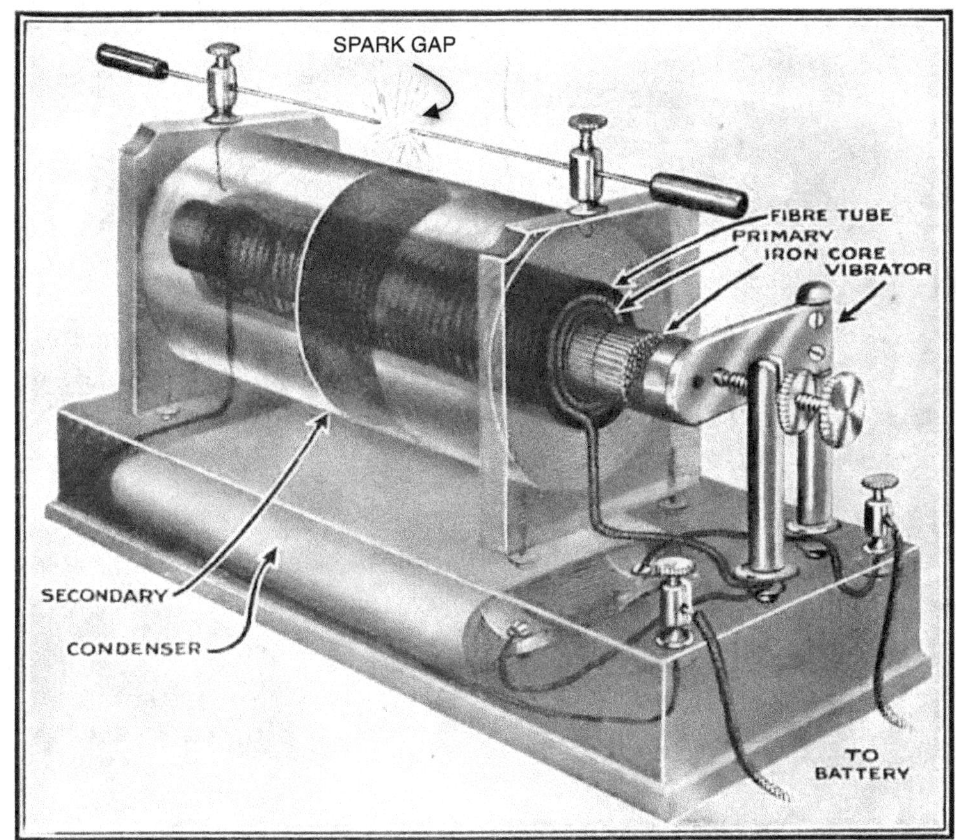

An induction coil, with the spark gap and the interrupter (here called "vibrator"); "spark gap" added.

THE *REPUBLIC-FLORIDA* DISASTER

The "Age of Electronics" received a healthy forward nudge when the Italian steamship Florida rammed the British liner *Republic* in a dense fog early Saturday morning January 23rd 1909 off the coast of North America, about sixty miles south-east of Nantucket lightship.

The outstanding features of this accident were:

The first use of wireless to summon assistance to a stricken ship at sea in imminent danger of sinking; the transfer of two thousand persons in open lifeboats over a rough midwinter sea without the loss of a single life; and the manner in which every important detail of the rescue work was published throughout the world simultaneously with its occurrence. The permanent result was compulsory laws requiring wireless on all ocean-going passenger vessels.[127]

At the time of the collision the *Republic* was bound from New York for Alexandria, Egypt, and the *Florida* was completing her voyage from Sicily to New York. The latter was crowded to capacity, mostly with survivors from the recent Messina earthquake. The *Republic* was on a straight course, laid from Sandy Hook a few hours previously, and the Florida was trying to pick up Nantucket Lights so that she could shape her course into New York harbor.

A dense fog had descended over the ocean only a few hours before the accident, and both vessels had slowed down. Each was blowing its foghorn seven seconds every minute. Each heard the other's warning blasts, and then when both horns became so loud that it was evident the two vessels were perilously close the captains exchanged signals by means of the foghorns. Yet the collision occurred!

From that moment onward the wild wastes of the Atlantic's Western fringe became a stage with all the civilized peoples of the World as audience. For the first time, they the World was able to follow the unfolding of the intense drama as it was happening. Wireless was the medium.

Aftermath of the Messina earthquake, December 28, 1908. The earthquake at 5:21 am was followed by a forty-foot tsunami that overwhelmed the coasts nearby. More than 100,000 people died.

Prior to sailing on the Republic, Binns had to meet the requirements for certification of "Proficiency in Radiotelegraphy" that had been decreed that year.

Certificate of Proficiency in Radiotelegraphy granted by the Postmaster-General.

This is to certify that, under the provisions of the Radiotelegraphic Convention, Mr. *John Robinson Binns* has been examined in Radiotelegraphy and has passed in—

(*a*) The adjustment of apparatus.

(*b*) Transmission and sound-reading at a speed of not less than 20 words a minute.

(*c*) Knowledge of the regulations applicable to the exchange of radio-telegraphic traffic.

The holder's practical knowledge of adjustment was tested on a *Marconi* set of apparatus.* His knowledge of other systems is shown below :—

It is also certified hereby that the holder has made a declaration that he will preserve the secrecy of correspondence.

Signature of examining officer _____________

R & Mackay for Secretary, G.P.O., London.

25 November 1908 (Date).

Signature of holder _____________

Date of Birth *16 September 1884* Place of Birth *Brigg, Lincs.*

*It is not intended to limit the employment of the holder to a particular system, but merely to indicate the particular system in which he was tested for adjustment of apparatus.

Q & S 1943 200/7/08—[1934] 200 10/08

To quote Willie Williamson, who looked at the Binns exam records held in the Liverpool Maritime Museum:

"He sat his exam on the 25th November 1908 at Marconi's depot at Seaforth in Liverpool. His sending speed was 26 wpm and he received at 29 wpm. His knowledge of "rules" was judged to be fair and his technical knowledge was good. He impressed the examiner who made a special mark that indicated that Binns was, quote, "a good man.""

Personal communication from Willie Williamson, Liverpool, UK.

The *Republic* was a twin screw vessel of 15,387 gross tons. She was built in Belfast, Ireland, in 1903 and commissioned the following year. She was 570 feet long, 67.8 feet beam and had a draft of 24 feet She was designed for Mediterranean cruises as well as the trans-Atlantic service so the majority of the accommodations were for first class passengers.

The Florida *at her launching in June 1905.*

The *Florida* was a 5,018 gross ton ship, with a length of 381.4ft and a beam of 48.1ft. She had twin screws that gave her a top speed of 14 knots. Because she was an immigrant ship—she only carried a handful of first class passengers, and no second class passengers—her interiors were no-frills, and built with an eye to cleanliness. The result was a ship with a distinctly modern air. However she had no wireless installation.

Between four and five o'clock in the morning 23rd of January these two vessels were so near each other that the foghorn of one was clearly audible on the other. Although both were proceeding with extreme caution they gradually closed in upon one another, and the sound of the foghorns grew correspondingly louder. There was nothing in this to excite alarm because the sirens of ships passing one another in a fog naturally increased in volume as they approach each other.

A little after five o'clock, however, the foghorns were much too strong for safety, and the two captains exchanged steering directions by means of their sirens. In the ordinary course of events, after the exchange of such signals, the two ships would have gradually drawn apart. In this particular case the unexpected happened—the *Florida* swung around as though on a pivot and bore straight down on the *Republic*.

Before anything could be done to prevent it, the Italian ship plunged her bow into the port-side of the British ship squarely amidships. Immediately after the collision the reactive forces of the impact cleared the *Florida* a few feet from the side of the *Republic* and she began to drift aft with the current. About ten feet of her bows still overhung the main deck of the British ship, and as the *Florida* dropped away toward the stern her overhang carried everything before it. The railing, bitts, boat davits, and stanchions on the port side of the *Republic* were broken, twisted or entirely carried away with as much ease as though they were made of thin wire rather than tough steel.

As the undamaged part of the bows slipped over the main deck promenade of the *Republic*, it penetrated some cabins, killing two passengers outright and seriously injuring two others. In the bows of the *Florida* four sailors were killed instantaneously.

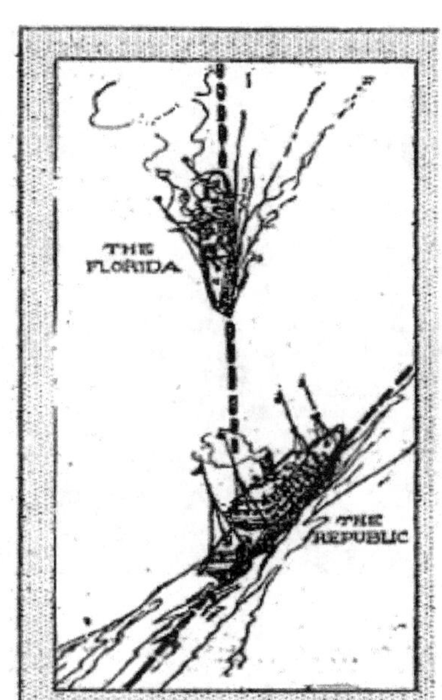

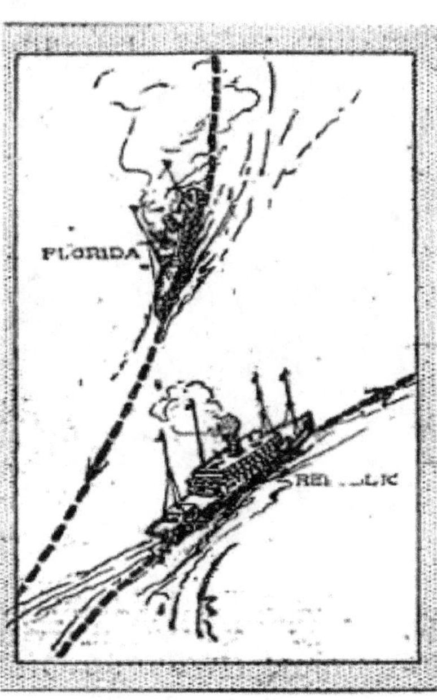

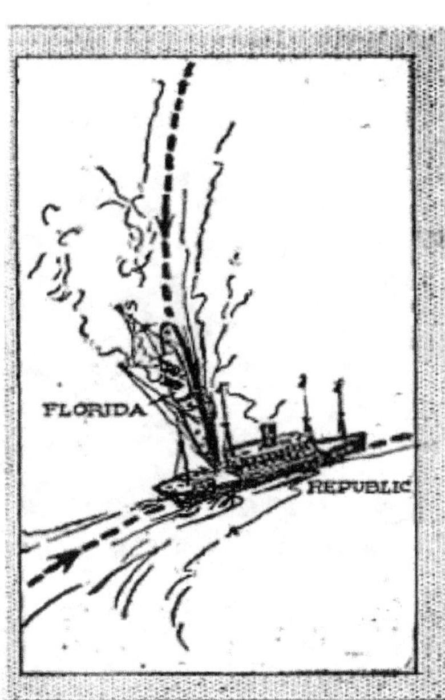

The collision, January 23rd, 1909.

The middle panel shows what should have happened as the two ships passed each other.
The right hand panel shows what actually happened.

As for the *Republic*, she had a jagged hole cut into her port-side reaching from the main deck downwards considerably below the water line. The hole was about fifteen feet across at its widest part.

The noise of the collision was deafening. It was something like the sharp crash that succeeds the discharge of a big naval gun, followed by a rumbling grinding noise not unlike that which accompanies a rolling earthquake. As the *Florida*'s bows were carrying everything away in her passage aft, a succession of bumps accompanied by a grinding noise told of the damage she was doing to the upper of the part of the main deck, which she left in a confused mass of broken steel strewn with wreckage of every kind.

While I stood on the threshold in the opening separating the two rooms in my cabin waiting for the crashing noise to finish, the port side and after walls of my cabin splintered up and fell in.

Portions of the boatdeck forming the roof of the cabin were also carried away. The telephone to the navigating bridge, which was fastened on the port wall of the cabin, was smashed to bits. Had I been seated at the normal operating position there is no doubt I would have been badly hurt in that mess.

Here's the wrecked deck of the Republic, with the carpenter and the carpenter's assistant holding the sounding rope used to calculate the rate at which the ship was sinking. The wireless cabin lies in the background, between the two figures; you can see its partially collapsed roof leaning across the wireless cabin. Note the two white shrouds on the left among the wreckage, and at right the line of gashes in the side of the aft saloons made as the bows of the Florida raked the deck.

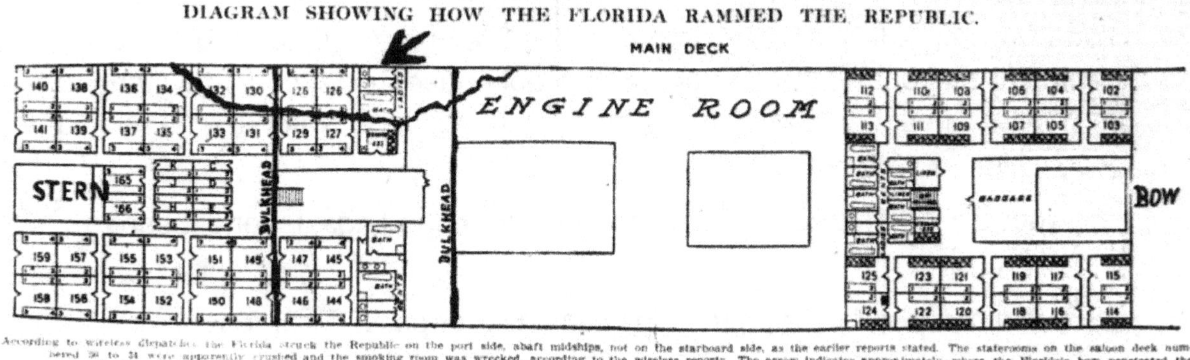

According to wireless dispatches the Florida struck the Republic on the port side, abaft midships, not on the starboard side, as the earlier reports stated. The staterooms on the saloon deck numbered 30 to 34 were apparently crushed and the smoking room was wrecked, according to the wireless reports. The arrow indicates approximately where the Florida's bow penetrated the side of the Republic. The black line shows the probable extent of the crushed plates and bulkheads. It was the breaking of these bulkheads that caused the sinking of the Republic.

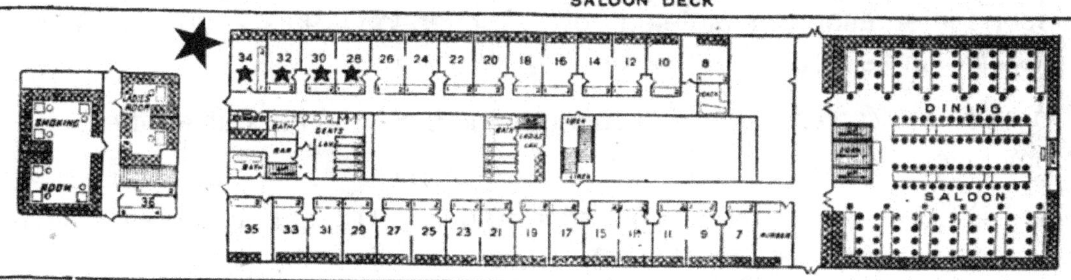

Here's the deck plan of the *Republic* showing the location of the damage. The stars show the cabins of the people in First Class who were injured or killed. Note the location of the engine room and the bulkhead to the stern of the engine room—the collision occurred squarely in the engine room, rather than at the location of the arrow in this drawing. The *Florida* recoiled, and then the overhanging portion of its bow swept across the saloon deck, raking the starred cabins and structures aft of these cabins, including the wireless cabin, located approximately at the large star. However *Florida*'s bow did not affect the main deck, so the passengers in the cabins by the main deck arrow were spared.

A Marconi wireless sending key, with a side lever that shifts the connection from sending to receiving and back. In the dark after the collision, Binns broke of the side level, so he had to hold it in place to be able to send messages, and lift it away to receive them.

A SHORT HISTORY OF CQD AND SOS

- Marconi adopted CQ in 1902 from the wired telegraph signal "CQ," meaning calling all stations—"Q" had been chosen because the letter would stand out—in English, the letter Q is normally followed by "U; a "Q" without a "U" would be distinctive;

- "D" added in 1904, because of minor misunderstandings; letter chosen to stand for "distress;"

- "SOS" introduced for German ships in 1905 -- does NOT represent any specific letters, but is easy to recognize;

- Second International Radiotelegraphic Convention, 1906, adopts SOS, to go into effect in 1908;

- *Not used by Marconi operators* because the U.S. had not ratified the Convention;

- The US Senate ratified the Convention April 3, 1912, to become effective May 25, 1912, after the sinking of the *Titanic*.

This newspaper photo shows Secretary of War William Howard Taft and his family returning to the U.S. on the *President Grant*, 1908.

On the voyage, Taft talked with Binns about SOS and the Convention—Binns was specifically assigned to the *President Grant* for this one voyage!.

Later, as president, Taft pushed for ratification of the Convention.

Did his conversation with Binns inspire Taft?

In the dark and cold, Binns reassembled his wireless apparatus, then…

…I made my way carefully through the wreckage to the bridge, to tell Captain Sealby that our wireless was working.

The moment I got back I began to send out the distress signal 'CQD' followed by the call letters of the *Republic* 'MKC'. The flashing of the spark in the inky darkness was the weirdest thing imaginable, and to the frightened steerage passengers who now crowded round my room, it must have had a most terrifying appearance.

Unfortunately I could not keep them away owing to the wrecked condition of my cabin. They could swarm around me, and their talking bothered me a great deal while I was listening for a reply to my calls. One of them sung out to me in a scared voice: "Oh! mister what is the matter?"

I told him to keep quiet, everything was alright, and they would soon be taken off the ship if they didn't make a noise. The crowd fell silent.

As near as I can remember I had sent out the distress call a couple of times when I heard Siasconset giving a general call. Apparently the operator had heard me, but couldn't make my signal out or who I was. Knowing how very weak my signals must be to him I sent very carefully and slowly:

"CQD! CQD! here is MKC MKC shipwrecked."

This I repeated three times, and to my great relief he replied by asking for particulars.

From his style of sending I recognized the "hand" of Jack Irwin who was on duty at the time.

Jack Irwin, the operator at the Marconi Station on Nantucket Island woke up feeling cold, so went over to out some coals on the fire, when he heard the weak signal Binns was transmitting:

CQD HERE MKC SHIPWRECKED

Thanks to his powerful apparatus he was able to pass the message on to ships in the vicinity and news of the event to the world.

While Binns was trying to get help, Captains Sealby of the *Republic* and Ruspini of the *Florida* determined that it would be safer for the *Republic* passengers to be transferred to the *Florida*—the *Republic* was the more damaged ship, clearly taking on water at the rate of a foot an hour.

As the fog lifted, the operation began. It took the sailors four hours of backbreaking work to transfer all the *Republic*'s passengers and much of the crew to the *Florida*.

While it was relatively easy for the passengers, many in light nightgowns and no shoes, to get from the *Republic* into the rowboats—the *Republic* put out stairs and a deck at water level—clambering up the twisting ladder on the side of the *Florida* as the rowboat moved up and down with the waves was another matter.

The transfer was just complete when the fog closed in again.

Here's the doughty crew who rowed all the Republic's *passengers and surplus crew to the* Florida, *then later rowed passengers and crew from both ships to the* Baltic.

Republic *crew photo taken in Genoa on the trip just prior to the disaster, when the* Republic *assisted the victims of the Messina earthquake.*

About the time the last transfer had been completed, I determined by the strength of their wireless signals that both the *Baltic* of the White Star Line, and *La Lorraine* of the French Line were within ten miles of the *Republic*. It seemed the *Baltic* was the nearest, so the Captains made arrangements via wireless to explode bombs at regular intervals.

On each ship the lookouts strained their ears listening for the report of the explosions, all to no avail. Occasionally some sailor would report he thought he heard the awaited sound. No chances were taken, consequently in each case careful bearings were taken of the direction he indicated, and steering directions sent by wireless to the *Baltic*.

How intense the concentration upon the *Baltic* was may be judged from just a few of the messages that were sent by us to her.

"*Baltic*—We can hear a bomb to the West of us. Is it you? *Republic*"

"*Baltic*. Steer North East at once. Sealby."

"*Baltic*. There is a bomb bearing North West from me. Keep firing. Sealby."

"*Baltic*. Can hear your whistle faintly you seem to be off our starboard bow. Sealby"

"B*altic*. Steer East South East. Listen for our bell. Sealby."

The story arrives in the press as it was happening, thanks to wireless! New York Evening World, *January 23, 1909.*

These are but a few of the many that were sent the space of one or two hours. Steering directions followed one another in rapid succession, so much so in fact, that when placed together as above they appear unintelligible and contradictory. Nevertheless they show quite clearly how the *Baltic* was really steered by Captain Sealby on the *Republic*'s bridge through the medium of wireless. The effect of these varying orders was offset by the fact that even when a wrong direction had been given it was impossible for the *Baltic* to proceed very far on a wrong course: owing to the decrease in the strength of her wireless signals I could tell her to check her course and transmitted new directions.

These apparently contradictory steering directions will give an idea of the zig-zag course pursued by the *Baltic* as she proceeded very slowly and with extreme caution in order not to run us down in the thick fog. In the end, she covered 200 miles in her search.

Everyone on board both the *Baltic* and the *Republic* were doing everything we could to get the former alongside before it became dark. Messages were flying across the intervening space in rapid succession, and although outwardly calm everyone was a working with feverish intensity, but all our devoted efforts proved inadequate. Darkness began to fall, completing the obscurity wherein we lay so helplessly. With the descending darkness our hopes began to fade because it appeared that what we had been unable to accomplish during daylight would be an utter impossibility in darkness.

The fog was still as thick as ever, and those on the *Republic* could neither see nor hear the *Florida*. The Italian ship was not equipped with wireless and it was impossible to learn her condition. She was as completely obscured from the sight or sound of all other vessels, as though she disappeared

Between five and six o'clock in the evening the outlook became bad. Darkness had descended upon the fog enshrouded scene, making it impossible to see more than a few feet. On the *Republic* the frequent soundings by the carpenter showed the water was still gaining in the holds at a steady rate of about a foot an hour, and at that rate the doomed ship could not last much longer.

It now appeared to be an impossible task to get the rescue ship alongside.

Despite the gloominess of the situation, no effort was relaxed. Bombs and rockets were being exploded as regularly intervals as before, and sound of the distant detonations were as eagerly awaited on each ship. An hour passed slowly by, but still the *Baltic*, though near, seemed to be as far off as she had been at noon.

At six o'clock in the evening we found we had but one bomb left, and enquiry elicited the same depressing information from the *Baltic*: she too had only one bomb left. I think from that moment everyone on the two ships who were aware of the exact state of affairs were convinced that the task of getting the *Baltic* to us was hopeless.

As soon as it was known the two ships had only one bomb left, arrangements were made to explode them at a stated time. I had to explain to the *Baltic* that our telephone to the bridge had been destroyed in the collision, so he would have to allow time for me to get to and from the bridge.

One of the first things we did was to check our chronometers by wireless. The *Baltic* arranged the time of his explosion in Greenwich Mean Time with his chronometer, and I hurried to the bridge with this information.

The *Republic* was to explode hers first, and then the *Baltic* was to follow at the appointed moment. A sufficient time was allowed between them for the purpose of communicating by wireless in case the first bomb was heard.

Shortly after *Republic*'s bomb went off, Tattersall, the operator on the *Baltic*, signaled back that the bomb explosion had not been heard on his ship.

I then went to the bridge. We had a brief consultation and the Captain told us to form a circle facing outwards, but in such a manner that we could still see quartermaster Crawley who was standing over the chronometer. At the exact second he raised his right hand, and we all strained our ears anxiously. The terrific concentration showed itself in the contracted muscles that distorted each man's face.

Absolute silence prevailed just as the exact second arrived.

Across the dark-curtained expanse of water the faint echo of a report sounded in my ears. Perhaps the sensitivity which wireless developed in me over the years enabled me to hear the faint sound. Mr J. R. Morrow, the fourth officer, standing next to me, said he thought he too had heard it. We both pointed toward the direction from which thought the sound came. A bearing was taken. I ran back to the wireless cabin and relayed the steering directions to the *Baltic*.

Each second of the succeeding hour was filled with anxiety and suppressed excitement. An explosion is generally the precursor of calamity but the explosion we had just heard was the probable forerunner of our rescue— at least that is what we hoped. The *Republic* was in very bad shape. About this time a fairly heavy swell was beginning to run, and she was moving very lumpishly with it. Besides this, the carpenter's frequent soundings were not very encouraging.

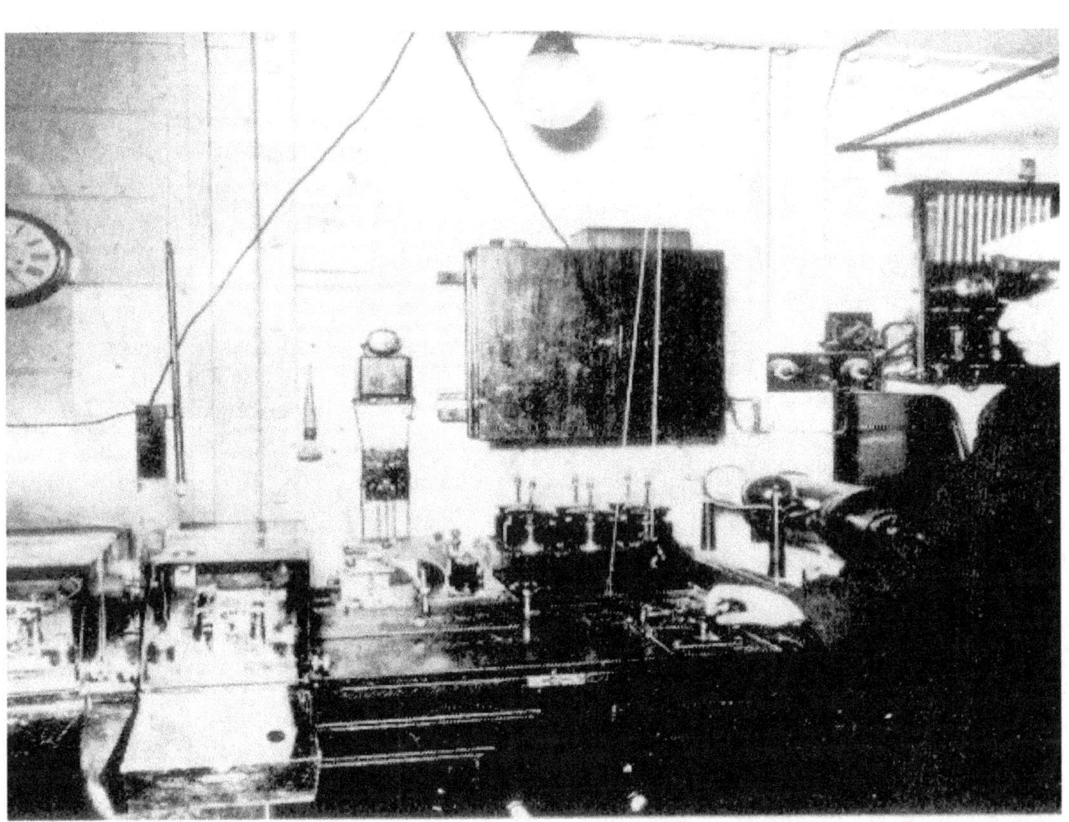

The radio room in the Baltic *with H. J. Tattersall at the key.*
Unlike the Republic, *the* Baltic *had a proper radio room built into her, equipped with direct contact with the bridge, so Tattersall could give Captain Ransom sailing directions according to the messages he received from Binns.*

I sat constantly in my cabin now. A sailor stood by to take any message that might come to the bridge. I dared not remove the head-phones.

About ten minutes after we had heard the report of the *Baltic's* last bomb, the wireless signals of that ship had perceptibly increased. I sent the welcome news to the bridge post haste. Captain Sealby sent down the reply telling me to inform the *Baltic* to continue on the same course on which he was, but to come with extreme caution.

During this time both ships were firing rockets at regular intervals in the hope they might be seen. An anxious half hour passed by during which the operator on the *Baltic* frequently asked if we heard anything of their foghorn, to which I had to continually reply in the negative, but still there was not the slightest doubt regarding the increasing strength of his signals. I told him so.

Shortly after darkness had descended upon us, the carpenter had brought me a small hand lantern; though its feeble light only illuminated its immediate vicinity, still with its aid I was able to hold the broken lever in place on the sending key with greater ease than I would otherwise have been able to. Each time I was sending the spark flashed out with its peculiarly weird bluish light.

Owing to the extreme darkness it was impossible to see the clock, so the times I am giving are only approximations. It would be about a quarter to seven when the strength of his signals had become so strong that I could hear the aerial from the receiving gear. This was an indication that he was at most one or two miles away from us.

I was in the midst of sending this information when the sailor came down with a message for Captain Ranson from our bridge conveying the information that we could hear the *Baltic's* foghorn from the same direction as the report of the bomb, and asking him to continue on the same course with great care.

The *Baltic* was sounding her foghorn automatically by electric control, one long blast lasting seven seconds every minute. Slowly the dull boom increased in strength to the listeners on the *Republic* as the oncoming ship closed in. By seven thirty the sound of the blast was so strong that Captain Sealby sent down a message which read:

> "Captain Ranson *Baltic*. Your foghorn very strong. You are too close for safety. Proceed slowly, Sealby."

How dangerous our own plight was can be judged from the repeated instructions to the *Baltic* to "come carefully." As the collision had occurred in our engine room and stoke-hold, the boilers had been put out of commission right away. This not only meant that the engines to our propellers were useless but also that the engine controlling the huge rudder was also inoperative, since there was no steam to make it work. As a consequence we were at the mercy of the sea and ever since early morning we had drifted about aimlessly wherever the current had carried us. Then too the stoppage of the dynamos had blackened out our powerful headlights. In their place were the comparatively feeble rays of our oil lamps. Under the circumstances any ship coming towards us could run into us before it was aware of our presence. It was for that reason the repeated admonitions to "come carefully" were sent to the ships around us.

But this message of apparent danger was to us also the harbinger of succor. So good were our prospects now, even though we were hidden by the fog, it appeared that nothing but the blackest disaster would prevent the *Baltic* from finding us.

A few more anxious moments passed by during which the *Baltic* was carefully feeling her way toward us, and then a message came down from the bridge, the contents of which practically settled our long but successful work. It read:

> "Captain Ranson. *Baltic.* You are very close now. Right abeam. Come carefully. You are on our port side. Have just seen your rocket. You are very close to us. Sealby."

Approximately at 7:30 pm while I was sending this message to the *Baltic*, a hearty cheer sounded over the crackling of my spark. As we only had a small portion of our crew on board—about forty all told—and they were scattered over various parts of the ship, I knew such cheering could not come from them. I left the key and went outside.

There in front of me was the *Baltic*—a blaze of light streaming from every porthole piercing the mist that hung around her.

It was the grandest sight that tired eyes ever saw!

The sailors' next job was to row all the passengers and surplus crew from the two disabled ships over to the *Baltic*. It was several hours of back-breaking work coming so soon after they had rowed all the *Republic*'s passengers over to the *Florida*, but this time the *Florida*'s crew could lend a hand. The transfer from the two disabled ships went smoothly—1522 people in all!

RMS Baltic ~ *the rescue ship.*

Binns went over to the *Baltic* for a quick nap. Then Captain Sealby, along with Binns and small crew returned to the *Republic*, with the hope of salvaging her, or at least having her sink in shallower waters. While the *New York* conveyed the *Florida* slowly back to New York harbor, yet another ship arrived on the scene, the *Furnessia*. She was assigned to help the the U.S. Revenue Cutter *Gresham* tow the *Republic* to shallow water.

Ships congregating around the Republic: *left to right, the* Lucania, *the* Republic, *and the* New York.

The stern of the *Republic* showing the after bridge with Captain Sealby, Mr. Crossland, Binns, and possibly another after they returned to the *Republic*.

The Republic *after the collision. The arrow, added, points to the location of the radio shack.*
A note pasted by Binns to the back of this photo says:
"The white patch amidships is the tarpaulin covering the gash made by the *Florida*.
Close examination shows the ruin caused by the *Florida*'s overhanging bows as she drifted sternward. All the stanchions, bitts, and railing were either carried away or twisted beyond recognition.
The wireless cabin just abaft of the mizzen mast [the third mast from the bow], had its port wall demolished. The one lifeboat still seen in its cradle was rendered useless when its davits were twisted beyond repair."

Despite all their best efforts throughout the day after the collision, the *Republic* took on more and more water and hardly moved. Finally it was clear that nothing more could be done. Binns and the crew were rowed over to the *Grisham*.

Captain Sealby and Second Officer Williams stayed aboard. At the last minute they leapt into the sea as the *Republic* slowly dipped under the waves stern first, cracking in half as she went down.

After many long harrowing minutes in the frigid water, first Sealby and then Williams were spotted and brought aboard the *Gresham*.

All in all, the only lives lost in the disaster were those seven who were killed by the collision itself!

Throughout Captain Ruspini of the *Florida* lived up to the noblest tradition of the sea. He was indeed a very brave and capable man. He never wavered, even though he knew that one of his crewmen killed in the collision was a fourteen year old cabin boy named Salvatore Amico, whose family—friends of Captain Ruspini—had been annihilated in the Messina disaster. The boy had pleaded with the Captain for the job. Now he was dead.

The Florida's *crumpled bow at the start of the repairs in the Brooklyn dry-dock.*

Ironically the *Florida* sank in the Mediterranean off Catalonia, rammed amidships by its convoy companion on a foggy December night in 1917 during World War I. She sank immediately, with considerable loss of life. She is now a site for amateur divers to explore.

When Sealby and Binns arrived in New York a huge crowd carried them bodily to the White Star offices, above.

The arrow (added) points to Sealby.

Reporters snapped pictures of them wherever they went. This photo shows Binns (left) with his pal Third Officer Stubbs.

The
CQD
medal
presented to the
crews of the
Republic, *the*
Florida, *and the*
Baltic *by the*
Republic's

THE RECOGNITION!

*New York Times Headline
January 27, 1909*

BINNS'S STORY OF WIRELESS WORK

Republic's Operator Gives The Times the First Account of His Long Vigil.

SIGNALS FROM BROKEN KEY

30

On their return to England, Binns and Tattersall received gold watches from Marconi at a celebratory dinner & concert in London. Photo shows Marconi handing this watch to Binns.

Postcard showing Luna park, mailed in July 1909, when Binns was starring in "Saved by Wireless."

In 1909 Marconi was in negotiations to spread the use of his system on American ships. Binns was the ideal person to be the "face" of Marconi: young, modest, hard-working, a man of the people and a hero.

Although Binns didn't want to be "made into a tin god," Marconi put pressure on him to participate in popular shows featuring wireless, and he gave in.

Among them were performances of a show called "Saved by Wireless" at Luna Park, the Disneyland of the day, complete with a shipwreck and Binns working the wireless apparatus.

Binns' performance as a wireless operator was, as you can imagine, true to life!

Binns was supposed to participate in the 3rd Annual Electrical Show in Madison Square Garden, sending wireless messages for attendees, in October 1909. It's not clear that he ever appeared. Marconi's efforts at getting contracts were succeeding, and he no longer needed the publicity. Binns was sent back out to sea.

Let Jack Binns Send a Wireless Message

Jack Binns, whose wireless message brought aid to the sinking *Republic*, operates the Wireless Telegraph

At the Electrical Show

See Hatching Chickens by Electricity—The Model Electric House—Electrical Control of Air Ships by Wireless—The Wireless Telephone—Electrical Laundry in Full Operation—All manner of novel and interesting manufacturing processes in full working operation.

A Veritable Fairyland of Beauty
A Wonderland of Electrical Marvels

Madison Square Garden.

Show Open from 10 A. M. to 11 P. M.

For personal examination of the various exhibits the morning or afternoon hours are suggested.

When the contracts Marconi was seeking were underway and he had no more use of Binns, he assigned Binns to the *Adriatic*.

On that ship Binns struck up a close friendship with Captain Smith, who was to go down with the *Titanic*.

Binns on the Adriatic, *1910*

*Captain E J Smith
of the* Adriatic, Olympic, *and* Titanic.

BINNS AND THE *TITANIC*

The Olympic leaving Belfast on her sea trial, May 1911.

When Smith was assigned to captain the *Titanic*'s sister ship, the *Olympic*, he chose Binns to go with him. But Ismay, the head of the White Star Line, was worried that Binns would remind people of the *Republic* disaster and bring bad publicity to his magnificent new ships.

When the *Olympic* arrived in Liverpool at the end of her sea trials, Binns received notice of his reassignment to less prestigious ships.

Binns was furious, but there was nothing he could do.

Meanwhile, there was no inquiry into the sinking of the *Republic*, and no real understanding of the importance of maintaining wireless channels open at all times, so no

lessons were learned — the sinking of the *Titanic* with its huge loss of life may well have been the result.

In 1910 Binns met his future wife, Alice MacNiff, in New Jersey, and began courting her when he was in the New York area..

By 1912 she declared that she wouldn't marry him if he didn't leave the sea.

He therefore sought a job as a journalist (he loved to write), and was offered a position as a reporter at the *New York Herald*. He started work on a Friday,

Just two days later, Sunday April 15,1912, the *Titanic* struck the iceberg and sank.

Alice Ann MacNiff at the time of her marriage to Jack Binns, 1914. She waited two more years after he became a newspaper man before saying "yes"—as she said, she enjoyed being courted!

The newspaper turned to Binns to report on the disaster, and to write about his friends Captain Smith and Jack Phillips, the one of the two Marconi men aboard the *Titanic* who drowned. The assignment saddened Binns greatly; he never knew what make of his good fortune in not sailing with the *Titanic*.

The newspaper editors wanted a scoop: firsthand accounts from the rescued passengers. Their idea was to send a tug out to meet up with the *Carpathia*, the ship that rescued the surviving *Titanic* passengers, with Binns on board to provide wireless communication.

The tug *Mary Scully* never fetched up with the rescue ship.

Here's Captain Sealby in his cap— he was no longer a captain. The lack of inquiry into the accident meant that he no longer could have a Master's ticket or captain a ship.

Nevertheless, Binns soon found himself the intermediary between the *Carpathia* and the shore, because that ship's very basic wireless system—the same system as the one Binns had used on the *Republic*—was unable to send messages the necessary distance, Therefore Binns relayed messages each way.

Among the messages were ones from Ismay to the White Star office in New York, demanding that the surviving crew of the *Titanic* be placed on the White Star liner *Cedric,* as soon as they were to land in New York City—the *Cedric* would then steam out of port and head to England. Ismay was concerned that a US investigation into the incident would prove to be yet another disaster for him and the White Star Line.

Setting up the wireless on the Mary Scully. *Binns is the person on the upper deck wearing a dark suit.*

Binns realized that if he divulged the content of the messages he would be subject to fines and even imprisonment, but at the same time he felt it was important to get to the bottom of why the disaster happened and to develop ways of avoiding similar catastrophes in the future. The fact that the *Republic* disaster had never been investigated lay heavily on his mind, together with the loss of his friend Captain Smith. The massive loss of life haunted him, especially as in 1910 he had written an article about the dangers of ships colliding with icebergs.

Binns and his colleagues on the tug devised a way to inform Senator Smith of Ismay's plans without revealing how they were discovered.

Smith hightailed it from Washington DC to New York, and boarded the *Carpathia*, subpoenas in hand.

Binns was called to testify at the inquiry headed by Senator Smith. Here are his words about his testimony:

> During the course of my examination Senator Smith asked my opinion on various matters relating to wireless operation, especially on the subject of wireless watches. noted that the operator of the *Californian* was the sole operator and had to get rest whenever the best opportunity arose—his was the only judgement as to the proper time to lay down the head-phones and turn in. I pointed out that the practice on the Atlantic Ocean was for operators to take their rest at night, but that it might be better to reverse the watches, with the most experienced operator on duty at night, and a person with a slight knowledge of Morse code and the wireless apparatus to be on watch during the day, to wake the chief wireless operator as necessary…. Continuing I expressed the opinion that there should be legislation making it mandatory for all ships to carry at least two operators—I had made the same recommendation in my official report to the Marconi Company after the *Republic* sinking.

In that Marconi report, Binns noted another problem he encountered in his effort to bring rescue ships alongside the *Republic*: interference by wireless amateurs chattering about the accident. At many points the signals from his weak apparatus were drowned out. A similar situation occurred when the *Titanic* was sinking, and then when the *Carpathia* was bring the rescued to safe harbor.

Further, the operators on the *Titanic* missed the warnings about icebergs because they were busy sending out paying messages, so didn't receive incoming ones until it was too late. In an article he wrote in 1910, Binns had expressed concern about icebergs and the need to have a way for warning about them to reach people in charge of the ship, but that article too went unheeded.

Binns always felt that the success of the Republic rescue was also a curse: it meant that people believed that all would be well in future disasters. It took the sinking of the Titanic, with its massive loss of life played out in the world's newspapers for everyone to see, for people to act to make wireless a means for averting future catastrophes.

J. Bruce Ismay Testifying at the Senate Inquiry.

X Indicates Mr. Ismay.

Photograph of English businessman J. Bruce Ismay testifying at a U.S. Senate Inquiry into sinking of the RMS Titanic, 1912.

JACK BINNS CHRONOLOGY

1884 ~ Jack Binns was born in Brigg, Lincolnshire, September 16th, 1884, and raised by his maternal grandmother and uncles in Peterborough, Northhamptonshire, where one uncle was a tailor of modest means, and the other worked on the railroad.

1898 ~ Binns passed his school leaving examinations and went to work at age 13 as a messenger for the Great Eastern Railway in Peterborough.

1898 ~ Binns' legs were nearly severed by a railway car. The doctors did not amputate because they believe he wild not survive. But survive he did! He took advantage of the year he spent in traction to read every book his doctors could bring him. Eventually he iwas able to walk on crutches and returned to work, this time as a junior operator. The wounds finally healed two years after the accident!

1901 ~ Offered the position of senior operator, second in charge, of the important Colchester Telegraph office, the main switching line between England, continental Europe, and the East, and the headquarters of the Eastern Command of the British Army. The teaught himself how to repair broken lines, apparatus, and connections, and devised ways to get around sunspot interference.

1903 ~ Passed tests and was accepted as an expert operator in the British Post Office Telegraph Service at Newmarket, the racing town.

1904 ~ Starting in December, Binns attended the Marconi training school in Seaforth, and was certified as a wireless operator after three months' training—the standard time was 6 to 10 months!

1905 ~ Received his first assignment, on the *Kaiser Wilhelm der Grosse* of the Hamburg Amerika Line. He sailed on German ships until 1908.

1908 ~ All foreign wireless operators were barred from working on German ships by order of the Reichstag, so Binns was sent, first to test new equipment for the Ostende-Dover ships, then to the Marconi Station at Crookhaven in Ireland for six months.

1908 ~ Assigned to the *RMS Republic.*

1909 ~ January 23rd, collision of the *Florida* with the *Republic.* Binns hailed as hero.

1909 - Spent most of year participating in events organized to promote the use of Marconi wireless systems.

1910 ~ Assigned to *Adriatic*, where his captain was E.J. Smith, later captain of the *Titanic.*

1911 ~ Due to be assigned to the *Olympic* and then the *Titanic* at the request of Captain Smith, but J. Bruce Ismay, head of the White Star Line, is worried that he might bring bad luck and bad publicity to these new ships. Therefore he is assigned first to the *Caronia* and then to the

Minnewaska, and also receives the post of Traveling Inspector, of Marconi systems created for him.

1912 ~ Resigned from the Marconi company to work for William Randolph Hearst at the *New York American*. Two days after he began his new career, the *Titanic* struck an iceberg and sank. He reported extensively on the disaster for the *New York American*. He testified at the *Titanic* inquiry.

1917 ~ Joined the Royal Flying Corps of Canada and then sent to join the Royal Air Force of Great Britain as a flight and wireless instructor.

1918 ~ Returned to newspaper work, and became radio and aviation editor of the *New York Tribune*, editing the weekly radio supplement. Also contributed articles to *Popular Science Monthly*, and wrote the prefaces to the *Radio Boys* Book Series

1924 ~ Joined Hazeltine Corporation as assistant treasurer. Hazeltine was formed to handle the patents developed by radio pioneer Louis Alan Hazeltine, and to develop and license new electronics patents for other inventors and corporations. Hazeltine developed the neutrodyne circuit that made radio commercially viable by decreasing interference and the high-pitched squeals of static.

1926 ~ Became treasurer of Hazeltine.

1927 ~ Became member of the Board of Directors of Hazeltine.

1935 ~ Became vice-president of Hazeltine.

1942 ~ Elected President of Hazeltine.

1952 ~ Elected Chairman of the Board of Hazeltine.

1955 ~ Resigned from the Presidency of Hazeltine.

1957 ~ Elected Honorary Chairman of the Board of Hazeltine.

1959 ~ Died of a stroke December 8th, 1959, in New York City.

Writings and Memberships

Starting in 1909 and even after beginning his work with Hazeltine, Binns had regular columns in a number of magazines, including *Popular Science* and *Collier's Weekly*. During his time as a newspaper reporter and thereafter he also wrote the Forewords to the thirteen books of the *Radio Boys* series published by Grosset and Dunlap between 1922 and 1930; Allen Chapman was the pseudonym used by the authors of this series.

Binns tried his hand at novel-writing himself, with *The Flying Buccaneer: A Novel of Adventure in the Skies* published by Nicholas L. Brown in 1923. This book was written to promote the advantages of heavier-than-air planes at a time when Zeppelins were becoming hugely popular.

Binns was a member of a number of professional organizations, including the Radio Club of America, the Society of Naval Engineers, the Armed Forces Communications and Electronics Association, and the Society of Silurians, a veteran newspapermen group.

SOURCES OF IMAGES

I: Collision diagram. Image found among Binns papers clipped from a contemporary newspaper.

iii: The R.M.S. *Republic*. Image from the Wikimedia Commons, Public Domain.

1: Jack Binns at about age 2: photo courtesy of Nicola Pike, great-granddaughter of Mary Ann Binns.

1: Edison with an early model of his phonograph. Image from *Stories of Great Inventors: Fulton, Whitney, Morse, Cooper, Edison*, by Hattie E. Macomber, Educational Publishing Company, Boston, 1897, presented by Project Gutenberg.

1: Binns family home at 207 Cromwell Road, Peterborough, from a photograph taken by Sidney Harbour, Peterborough, UK, October 11, 2009.

2: The front of the Great Eastern Railway Station, Peterborough, around 1910. Image from an old postcard, courtesy of John Alsop of the John Alsop Collection.

2: A view of the marshaling yards, Great Eastern Railway, Peterborough. Image from an old postcard, courtesy of John Alsop of the John Alsop Collection.

3.: Splinting of leg fractures. Illustration from *The Hodgen Wire Cradle Extension Splint: the Exemplification of this Splint with Other Helpful Appliances in the Treatment of Fractures and Wounds of the Extremities and its Application in Both Civil and War Practice*, by Frank G. Nifong, C.V. Mosby Company, St. Louis, 1918, Google Books, Public Domain.

4: Connections of the Eastern Telegraph Company, 1901. Wikimedia Commons, Public Domain.

5: Young Binns in his Marconi cap. Personal photo.

6: S.S. *Kaiser Wilhelm der Grosse*, c. 1897. Photograph by A. Loeffler, Tompkinsville, NY, United States Library of Congress Prints and Photographs Division, LC-USZ62-69220.

6; S.S. *Prinzessin Viktoria Luise*, 1905. Image from http://www.gjenvick.com/HamburgAmerikaLinie/1905-AcrossTheAtlantic-Photos-PrinzessinVictoriaLuise.html.

6: S.S. *Grosser Kürfurst*. Hamburg-Amerika Line postcard mailed in 1905.

7: "Selling the daily newspaper on board the Hamburg-Amerika Liner *Amerika*." Image from the *New York Tribune*, October 7th, 1906, in the United States Library of Congress, Chronicling America Project.

7: Back of a postcard mailed from the *Blücher* at Spitsbergen, 13 July, 1906. Image courtesy of Kent Larsson, who owns the postcard.

8: Guglielmo Marconi, circa 1903. Image from the United States Library of Congress, photoprint by Falk, LC-USZ62-77563.

8: The Marconi cabin on the deck of a ship. Photo published in the *New York Evening World*, January 25, 1909, United States Library of Congress, Chronicling America Project.

10: Inductively coupled coherer radio receiver circuit from Guglielmo Marconi's April 26, 1900 patent. Image from Wikimedia Commons, Public Domain.

11: An induction coil, with the spark gap and the interrupter (here called "vibrator"); "spark gap" added. Image from *The How and Why of Radio Apparatus*, by Harry Winfield Secor, 1st Ed., New York, Experimenter Publishing Co., 1920, p.5, as available in the Wikimedia Commons, Public Domain..

12: After the Messina earthquake. Photo by Underwood & Underwood, 1908, in the Wikimedia Commons, Public Domain.

13: Binns' Certificate of Proficiency in Radiotelegraphy. Image scanned from the original certificate.

15: Collision diagram. Image found among Binns papers clipped from a contemporary newspaper.

16: A view of the wrecked deck. Photo taken by Captain Sealby's steward, A. Lax; published in several contemporary newspapers, Public Domain.

17: Deck plan of the *Republic* showing the location of the damage where the *Florida* rammed the *Republic*. Image from the *New York Tribune*, January 25, 1909; United States Library of Congress, Chronicling America Project.

17: Marconi wireless key with side lever. Image courtesy Tom Perera, W1TP Telegraph & Scientific Instrument Museums: http://w1tp.com.

18: Photo of the Taft family arriving in New York. Image from the *New York Tribune*, December 21st, 1907, from the United States Library of Congress, Chronicling America Project.

19: Jack Irwin. Image from *The Aerial Age: A Thousand Miles by Airship Over the Atlantic Ocean; Airship voyages over the Polar Sea; The Past, the Present and the Future of Aerial Navigation*, by W. Wellman. New York, A. R. Keller & Company, 1911.

20: Reconstruction of the clamber up the ladder to the *Florida*. Image from the *New York Evening World*, January 25, 1909; United States Library of Congress, Chronicling America Project.

20: *Republic* crew. Photo provided by Paul O'Brien—his ancestor Thaddeus Crawley, Quartermaster,[1] is just below the man in front of the White Star.

21: Headlines in the *The New York Evening World*, January 23, 1909. Image from the United States Library of Congress, Chronicling America Project.

23: Radio room on the *Baltic* with Henry Jack Tattersall (?) at the key. Photo in several contemporary newspapers.

25: R.M.S. *Baltic*. Image from a postcard , Wikimedia Commons, Public Domain.

25: Ships congregating around the *Republic*. Original photo in Binns' papers.

26: The stern of the *Republic* showing the after bridge with Captain Sealby, Mr. Crossland, Binns and possibly another. Photo taken from the *Baltic* as she left the *Republic* to return to New York City by Miss Ruth McAfee, a passenger from the *Republic* on the *Baltic*, and printed in the *New York Tribune*, January 26, 1909, United States Library of Congress, Chronicling America Project.

26: The *Republic* after the collision. Binns' personal copy of the photo taken from the *Baltic* by *Republic* passenger Ruth McAfee.

27: The *Baltic* returns to New York with survivors. Image from the *New York Evening World*, January 25, 1909; United States Library of Congress, Chronicling America Project.

28: Captain Ruspini. Image in the *Marion [Ohio] Daily Mirror*, February 18, 1909, United States Library of Congress, Chronicling America Project.

---

[1] A quartermaster is a ships' officer reponsible for signals and steering.

28: The crumpled bow of the *Florida*, in dry dock in Brooklyn. Image from contemporary newspapers, source unknown.

29: Crowd greeting the *Republic* crew at the White Star office, arrow added. Image from the *New York Tribune*, January 27, 1909; United States Library of Congress, Chronicling America Project.

29: Third Officer and Binns' pal Stubbs (left), with Binns on the street in New York after the *Republic's* sinking. Image from the *New York Tribune*, January 27, 1909, United States Library of Congress, Chronicling America Project.

30: Medals received by Binns, currently in the Peterborough Museum. Image provided by Sidney Harbour and printed here courtesy of the Peterborough Museum and Art Gallery, Great Britain.

31: Marconi presenting watch to Binns. Personal photo.

31: Program, photographed from original object by VUL.

31: Watch presented to Binns, currently in the Peterborough Museum. Image provided by Sidney Harbour and printed here courtesy of the Peterborough Museum and Art Gallery, Great Britain.

32: Luna Park. Postcard mailed in 1909 and sold on eBay.

32: Ad for the Electrical Show, *New York Tribune*, October 16, 1909. Image from the United States Library of Congress, Chronicling America Project.

33: Jack Binns on the Adriatic, summer 1910. Personal photo.

33. Captain Edward John Smith, RD, RNR (1850 – 1912). Photo from *Wreck and Sinking of the Titanic*, by Marshall Everett, Chicago, L.H. Walter, Publisher, 1912.

34. The *Olympic* leaving Belfast. Image from the George Grantham Bain Collection, United States Library of Congress Prints & Photographs Division, LC-USZ62-76281.

35. Installing the wireless on the way to meet the *Carpathia*. Image from the George Grantham Bain Collection, United States Library of Congress Prints & Photographs Division, LC-DIG-ggbain-10350.

37. Ismay testifying at the Senate Inquiry. Image from the New York Times, April 20, 1912, Wikimedia Commons, Public Domain.

ACKNOWLEDGEMENTS

The number of people I must acknowledge for their help in making this book possible is boundless. However, I must confine myself—for those of you who go unmentioned, my gratitude and apologies.

No more enthusiastic crowd can be imagined than those who organized the events around the commemoration of the *Republic- Florida* collision on January 23rd, 2009. From the moment John Allen contacted me in 2008 from Scunthorpe UK, a town near Brigg, my grandfather's birthplace, to the moment of this writing, I have been supported by interested and helpful people who have been able to flesh out many details of this story.

In addition to John Allen, who started the ball rolling and kept it going, I am so grateful to (in no particular order):

Sidney Harbour of Peterborough, who provided me with fresh photos of the Binns medals and scroll, took the time and effort to photograph the Binns home in Peterborough, and found and scanned pictures of the Peterborough Railway Station, among many other gifts;

Andrew Longworth, Collections and Access Assistant at the Peterborough (UK) Museum and Art Gallery, for his help with the photographs of the Binns medal;

David Barlow of the Lizard, who prodded me to develop information about the accident, its preludes and aftermaths, and transformed this information into fascinating articles; furthermore, David's interest in Jack Irwin prompted me to dig out the information about him that my grandfather just hinted at—I discovered that Irwin is worthy of a book just for himself;

Willie Williamson at Liverpool, who among many other contributions, looked up Binns' examination records and those of his colleagues Tattersall and Balfour at the Marconi School at Seaforth;

John Alsop of the John Alsop Collection of old postcards, who so kindly allowed me to reproduce images of these cards here;

John Bowen of the Chelmsford Amateur Radio Society, who opened up the Marconi wireless station at Chelmsford for the commemorations, and went on the search for information about Chelmsford man Tattersall, the head Marconi operator of the Baltic;

Barbara Dougan of the U.S. National Park Service, a wireless enthusiast and saver of lives, who was instrumental in organizing the 100th anniversary commemoration at Cape Cod, and who introduced me to the Marconi history at Cape Cod;

Tom Perera, whose website (http://w1tp.com/) was my introduction into the mysteries of wireless equipment and equipment collection; his invitation to give a talk at the annual meeting of the Antique Wireless Association in August 2009 was further impetus to completing this book; at this meeting I had the great pleasure of speaking with a most impressive group of people, husbands and wives together, for whom the history of wireless communication is a deep passion;

Tom also introduced me to Russ Kleiman, to whom I am indebted for the identification of the elements of the *Republic*'s Marconi set-up;

Kent Larsson, with whom I exchanged a fascinating discussion about early radio transmissions in the Arctic;

my cousin Nicola Pike, who provided early pictures of the Binns family, and whose enthusiasm for all things Binns spurred me to complete this book.

I can't forget my family, who looked over earlier versions of this book with their sharp eyes, and suggested many improvements.

The Internet has been a faithful collaborator in this effort, making it possible to gather the people, the information, and the images that I hope make this book interesting. Through the 'net I could find nearly everything I needed, from the history of wireless to the history of political and economic events at the beginning of the 20th Century.

Among the websites that I have found particularly useful are:

"Early Radio History," created by Thomas H. White (http:/ /earlyradiohistory.us/), which contains some of the very earliest documents concerning wireless history;

the sites of the United States Library of Congress; the collection of contemporary newspapers of the *Chronicling America Project* has been particularly useful;

the *New York Times* archives: for articles detailing of any number of issues, especially in the political realm—great reporting, then and now!

Googlebooks, which has provided easy access to books that would be otherwise unavailable to me;

and of course Wikipedia and the Wikimedia Commons, which have guided me to original literature both on the web and off, and provided numerous illustrations.

Speaking of illustrations: with the exception of those images for which I have obtained permission, I believe that all are in the public domain. If there is an image for which I should have obtained permission, but didn't, please, dear reader, let me know and I will make amends.

Any errors and omissions in this book are, of course, mine alone. I will be deeply grateful to those who point them out to me.